OPENSCAD FOR 3D PRINTING

By Al Williams

Copyright © 2014 by Al Williams

All Rights Reserved. Except as permitted under the United States Copyright Act of 1976, no part of this publication may be reproduced or distributed in any form or by any means, or stored in a data base or retrieval system, without the prior written permission of the author.

ISBN: 1500582476

ISBN-13: 978-1500582470

Dedication

Always, to Pat

Table of Contents

Chapter 1. OpenSCAD and 3D Printing 1
    Cloud Computing 5
    In Summary 5

Chapter 2. OpenSCAD Basics 7
    Parameters 10
    Comments 11
    Rectangular Prisms and Cones 12
    The Polyhedron 15
    In Summary 16

Chapter 3. Operations 17
    Differencing 20
    Scaling and Resizing 22
    Mirroring 22
    Intersection 23
    Math and Variables 24
    Printing 27
    A Simple Panel 29
    Summary 33

Chapter 4. Two Dimensions and Extrusions. 35
    Extruding Imports 38
    Copying a Flat Solid Object 38
    GIMP 39
    Inkscape 41
    OpenSCAD 42

> Print! 44
> Wrap Up 45

Chapter 5. Programming OpenSCAD 47
> Variables 47
> Modules and Functions 47
> For Loops 50
> Thingiverse Customizer 57
> Wrap Up 59

Chapter 6. Libraries 61
> Some Predefined Libraries 62
> Wrap Up 65

Chapter 7. Advanced Topics 67
> Strings 67
> Importing STL 67
> Advanced Transforms 68
> Projection 70
> Wrap Up and Summary 71

Thanks 73

Introduction

When I was in school, many years ago now, I had a few semesters of drafting. Being progressive, we had about a week of "computer aided drafting." In those days, that meant punching a few cards that said something like "RECT 10 10 5 5" and handing it to a computer operator. Later in the day you came back and picked up a big curly piece of paper that had a rectangle on it. I was less than impressed.

I'm primarily an engineer, not a draftsman, and as the state of the art in Computer Aided Drafting (CAD) advanced, it seemed like my prediction was correct: no one wanted to write cryptic commands to drive drawing. They wanted to use a mouse or trackball and draw on the screen as if it were paper.

At about the same time, things were changing in the electronics industry. When I worked for a major semiconductor manufacturer, our computer chip designs were on huge rolls of paper that had been meticulously drawn. As computer design moved from table tops to computer workstations, it seemed obvious that you wanted tools that let you draw those same huge rolls of paper virtually.

Obvious, perhaps, but also wrong. Drawing computer circuits gave way to a much better system that used computer-like languages to describe circuitry. Instead of drawing 96 interconnecting lines, you could write something like ramdata=ram_cs?data[31:0]:32'bz;

If you don't know the language, that may seem cryptic but it is compact, succinct, and—most important—if I decide to change the bit width to 16 or 64 or 256 it takes just a second to make that change.

A lot of drafting has gone the same way. There are tools that let you draw as though your computer screen is a piece of paper. That's not OpenSCAD. OpenSCAD is the great grandson of the old CAD punchcard system I used in school. Of course, there are no cards, it is less cryptic, and it creates 3D objects with relative ease.

Just as computer design moved away from drawing, even the big drawing-style CAD tools often some way to do parametric drafting. OpenSCAD is the extreme. You draw nothing. You describe everything. It shows you a picture of what you described so you can refine your description.

This book will teach you the fundamentals of using OpenSCAD to develop 3D models for printing. You can do a lot of things with OpenSCAD, but I'll only focus on the parts that help you get output from your printer.

I won't talk much about the rules of what makes an object good to print. If you are interested in that, you might want to read one of my other books, *Understanding 3D Printing*. However, the exact nature of what you can do and can't do varies by the type of printer and material you are using. In general, though, an object needs to have a flat base to sit on the print bed, and there are usually limits to how much material can hang over an edge or cross over a void.

OpenSCAD isn't the best choice for all kinds of 3D modeling. It excels as what I think of as engineering models. Things like front panels, bearings, shims—things that have regular shapes that you might draw in a drafting class. It is not so good at modeling things that have irregular surfaces like animals or plants or people. You might be able to model some things like that in OpenSCAD, but it would be difficult and frustrating compared to using a program that is made to model artistic objects (like, Blender, for example).

Inside this book, you'll find easy-to-digest Chapters that will help you get the most out of 3D drafting with OpenSCAD. If you haven't used OpenSCAD, you should probably work through Chapters 1 to 4 in order. The remaining chapters cover more advanced techniques that you could revisit as you need them in any order (more or less).

It is tempting to pad out a book like this with lots of graphics of 3D models. You'll see some in these pages where they help illustrate a point. However, you should really read this book with OpenSCAD open on your desktop. Type in or copy and paste the example code into OpenSCAD and see the output for yourself. Think about a change and try to make it work. You'll learn more that way than just looking at screenshots on the page of the book.

Chapter 1. OpenSCAD and 3D Printing

You can hardly read a newspaper lately without some story appearing about 3D printing. I'm going to assume that if you are reading this book, you are familiar with 3D printing (and, if you aren't, you might want to check out one of my other books, Understanding 3D Printing, available on Amazon).

The quick summary, though, is that you start with a 3D model (usually in an STL file). The flow in Figure 1.1 shows this in the second row (it assumes you created an STL file in the first row). You run that through a "slicing" program that cuts the object (conceptually) into thin slices and plots the movement of the printer head for each slice. This outputs a G code file. That file will drive the printer to actually make the object. If you don't own a printer, you might send the model file to a service company to have it printed for you.

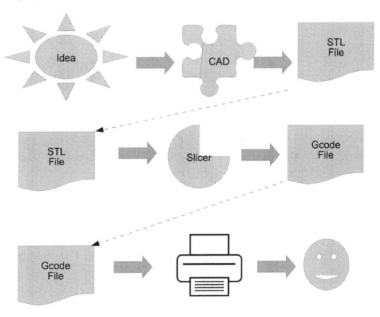

Figure 1-1. 3D Printing Workflow

The point to this book is how do you get that original model file? In other words, how do you go from idea to an STL file (the top row in Figure 1-1). Like most things, there are several possible answers. First, you could download a model from a Web site (like Thingiverse, for example).

That's handy because you don't have to do anything but download it, but it also means you can't get exactly what you want unless you just happen to get lucky.

Another answer is to use a 3D scanner to take a "picture" (so to speak) of a real world object. Scanners aren't very common these days, although there are a few out there. The results are not always that great, either. Plus, sometimes you don't have a real object to start with—you want to create from scratch.

The third way to get an STL file is to create it using a Computer Aided Design (CAD) program or a 3D modeling program. When you think of a CAD program, you usually think of a graphical program that uses the mouse to draw. There are also 3D modeling programs (like Blender) that are more suited for artistic uses. You can certainly use these tools, and many do. However, for complicated drawings it can be painful to use these tools and even more painful to make changes.

Think of an example. You are building an electronic project and you need a front panel with 16 holes for lights and switches. The holes need to line up just right. Most programs will have some kind of alignment tool, of course, but you need to get all the holes in exactly the right spot and lined up by pointing and clicking, using grids, and alignment tools. What if you have to change something? Adding a hole may mean moving a bunch of holes and redoing everything again.

Another form of CAD program is a parametric CAD program (like OpenSCAD). The idea is that instead of drawing a physical shape that describes what you want to build, you actually use a special language to describe the object. This is very similar to a computer program and can have variables and rules. That means I can make changes and things will work out automatically.

Consider that I want to make a cube with a hole exactly 25 millimeters from the edge of the cube and one in the exact center. Instead of drawing it I could describe it with this simple OpenSCAD program:

```
cubesize=125;   // size of cube

offset=25;      // location of one hole

holesize=3;     // size of holes
```

```
difference() {
   cube(cubesize);
// punch hole at offset
      translate([offset,offset,0])
 cylinder(r=holesize,h=cubesize*2);
// punch hole at center
      translate([cubesize/2,cubesize/2,0])
         cylinder(r=holesize,h=cubesize*2);
}
```

To make OpenSCAD code stand out, it will appear between lines like the fragment above.

You can see the resulting object in Figure 1-2. Don't worry about the details, yet, but you can see that all the sizes are set at the very top of the program. The locations of the holes are expressed mathematically using those variables (for example, cubesize/2). To change the size of the cube, you simply change the variable and run the program again. The hole will be centered automatically.

This is just a small taste of what you can do with OpenSCAD. The key is that you can automatically place and size features of an object using rules instead of hand drawing each one. It is worth noting that some traditional CAD programs have some ability to do this in addition to the graphical user interface. OpenSCAD doesn't take this hybrid approach. You write the program, view the output, and when you are ready you generate the output file that will go to your 3D printer.

OpenSCAD can do a lot more. For example, Figure 1-3 shows one of the examples that ships with the program. You can see threads, gear cuts, spirals, and more all created by an OpenSCAD program.

Figure 1-2. A Simple Open SCAD Example

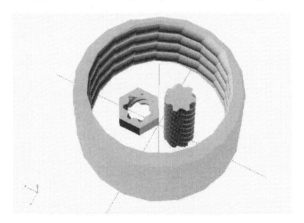

Figure 1-3. A More Complex Example

    The first thing you need to do before proceeding to Chapter 2 is install OpenSCAD. The software is free and available at http://www.openscad.org. You can download versions for Windows, Linux, BSD Unix, and the Mac. Follow the installation instructions on the Web site and when you run the program you should see a window that looks like Figure 1-4.

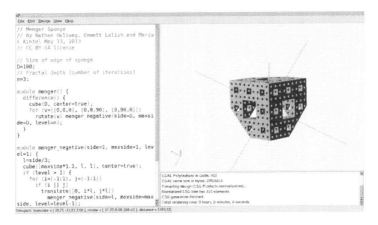

Figure 1-4. The OpenSCAD Interface

Cloud Computing

If you don't want to install software on your computer, there are at least two "cloud-based" versions of OpenSCAD at the time of this writing. These don't work exactly like a desktop version of OpenSCAD, but they will work with most of the examples in this book. Find one at http://www.openscad.net and the other at http://www.fabfabbers.com/openscad/. You need a fast connection and modern browser to make these Web apps work.

In Summary

In this chapter, you've learned why you might want to use OpenSCAD to generate 3D printing models. Although it isn't point and click like a traditional CAD program, the use of parametric modeling is very powerful and can create useful and maintainable results without a lot of effort.

Chapter 2. OpenSCAD Basics

When you start OpenSCAD (regardless of which operating system you are using) you should see a screen like the one in Figure 2-1 (your menu may look different and you may have different colors, but the basic layout should be the same). The pane to the left is a text editor. The right pane is split in two pieces. The top portion is the drawing space where your new creation will appear. The bottom part is where informational messages will appear (including errors). At the very bottom of the screen there is a status bar that has more dynamic information that can be helpful.

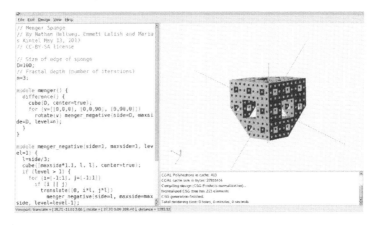

Figure 2-1. The OpenSCAD Interface

From the File menu, select examples and then pick example001.scad. Assuming they don't change the examples, you'll see the following code in the editor pane:

```
module example001()

{
        function r_from_dia(d) = d / 2;


        module rotcy(rot, r, h) {
                rotate(90, rot)
```

7

```
            cylinder(r = r, h = h, center = true);
    }


    difference() {

        sphere(r = r_from_dia(size));

        rotcy([0, 0, 0], cy_r, cy_h);

        rotcy([1, 0, 0], cy_r, cy_h);

        rotcy([0, 1, 0], cy_r, cy_h);

    }


    size = 50;

    hole = 25;


    cy_r = r_from_dia(hole);

    cy_h = r_from_dia(size * 2.5);
}


example001();
```

If you press F5 (or pick Design | Compile from the menu), you should see the result in Figure 2-2.

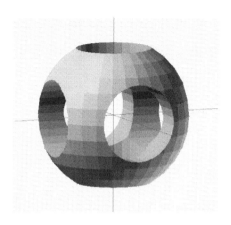

Figure 2-2. Example Drawing

This example uses a lot of things we won't learn about until later chapters, but it is a good test case to see that everything is working. If you take your mouse and drag around the display area (the top right pane of the program) you'll see you can manipulate the view. Click and drag rotates the view (press shift to constrain the rotation to two axes). Scrolling with a scroll wheel zooms in and out (or you can drag with the middle mouse button or Shift+right drag). Right click dragging pans the display.

Other useful keystrokes are Control+0 to reset rotation and Control+P to reset movement. The Control+P shortcut is especially useful since rotation occurs around the center of the viewport, not the origin, so if your rotation seems crazy, try a Control+P to get back to the center. You can also use the View menu to control what and how you see your object. The camera directions (top, bottom, front, diagonal, etc.) are handy when you've confused yourself dragging the mouse.

A few of the menu items (like Animate) won't make sense until you learn more about OpenSCAD. The Perspective view (the default) makes distances far away from the virtual camera seems shorter than distances closer (as they appear in real life). In Orthogonal view, things are drawn with their correct length. Combined with the view directions like top and front, this allows openSCAD to produce output similar to a traditional 2D engineering drawing of a 3D object.

To see the difference, consider this pair of long bars:

```
cube([10,2000,10]);

translate([20,0,0]) cube([10,2000,10]);
```

If you like, you can enter this into the OpenSCAD editor (remove any other text that is there or use File | New on the menu). Figures 2-3 and 2-4 show the output in Perspective mode (Figure 2-3) and Orthogonal mode (Figure 2-4). Notice in Figure 2-3 the bars angle towards an imaginary vanishing point on the horizon. In Figure 2-4 the bars maintain their true shape and proportion.

Figure 2-3. Perspective View

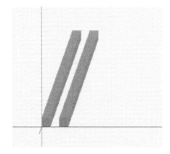

Figure 2-4. Orthogonal View

Parameters

Start with a blank editor pane in OpenSCAD and enter the following line:

```
cube(5);
```

Then press F5 to render the output. You should see a small cube. Zoom in (use the scroll wheel of your mouse) until you can see it comfortably. You might also want to rotate around it to see it from different angles.

Note that like most computer languages, OpenSCAD is a bit picky about syntax. If you leave the semicolon off, for example, you'll see an error message down in the console pane. You must use the same capitalization and the right pair of brackets, as well.

Not surprisingly, the cube is 5 units square and it has a corner at the origin (if you can't see the axis lines, pick View | Axes from the menu). What if you wanted it centered? Try this:

```
cube(size=5,center=true);
```

Press F5 again and there you are. In theory, you don't have to put the size= and center= here. That is "cube(5,true);" would also work. However, if you have a lot of different parameters, using their names makes sure you have them right and also helps you read your code later and understand what it is trying to do. Just as an example, using "true,5" doesn't work (well, it sort of works, but it doesn't do what you think at all) but using "center=true,size=5" will work.

As you'd imagine, these things aren't specific to the cube. Later, you'll see other shapes and even things that aren't exactly shapes that allow you to name parameters to them in the same way.

Comments

Speaking of understanding your code later, comments are always a good idea. You can start a comment line with //. That kind of comment goes to the end of the line. You can also bracket a comment with /* and */ that can span lines or live inside of a line. For example:

```
// My first cube

cube(center=true,
   /* need a small one here */ size=5);
```

White space and blank lines are usually unimportant so you can use them to format your code as well. So while it isn't very pretty, you could write:

```
cube(

center=

true,

size

=5);
```

That's an extreme example, but it does show that spaces don't mean much (unless they are inside quotation marks).

Rectangular Prisms and Cones

Sometimes you want a cube that isn't the same size on each side (technically, that's not a cube, but to OpenSCAD it is). You can provide a triplet of numbers to the size argument that specify the size for the X, Y, and Z axis. Try this:

```
cube(size=[10,15,25]);
```

Don't forget to press F5 to redraw. You might wonder what the units of these numbers are. That's a good question and the answer is: any units you want them to be. OpenSCAD (and STL files, for that matter) are unitless. The slicing software will interpret the units in some way. Nearly all applications I've seen use millimeters, so if that's true of your software the cube is 10 millimeters by 15 millimeters by 25 millimeters. But as far as OpenSCAD is concerned it might be feet, miles, or light years.

Examples often use whole number units, by the way, but OpenSCAD has no problem with decimal numbers. Try changing the 25 in the example to 12.5. OpenSCAD can also do math so you could replace the 25 with 25/2 to get the same result as 12.5. This is sometimes useful to show a proportion, as in:

```
cube(size=[10,2*10,5]);    // cube twice as deep as it is wide
```

This is especially useful if you use variables. That means you could write:

```
cubeface=10;

cubethickness=5;

cube(size=[cubeface,2*cubeface,cubethickness]);
```

A variable is just a name for number. They work a little different than in some popular programming languages but for now, you can treat them as simple variables—anywhere you see cubeface you can just as well say 10. Of course, if you wanted to change the size, it is much easier to just change the variable instead of having to change it everywhere in the code (especially if you were making hundreds of boxes).

Just like a cube can also make rectangular prisms, the cylinder command can create both cylinders and cones. Try this (don't forget the F5):

```
cylinder(h=10,r=15);
```

You might notice that you can see the cylinder produced isn't really smooth. It is made up of a certain number of slices and you can easily see that on the screen. The reason is that OpenSCAD purposely picks a low number of slices to make drawing faster. You can set a higher value to get better looking cylinders (and circles) but with slower drawing using the built in $fn variable. You definitely don't want to print anything circular with the default number of slices.

You can update the value of $fn for a single object or for all objects. If you just need one object to use more slices, you can add a $fn value to the parameter list:

```
cylinder(h=10,r=15,$fn=100);
```

If you draw that line of code, you'll see a very smooth surface around the cylinder. You can also set $fn globally:

```
$fn=100;
```

The $fn special variable works with $fa and $fs when OpenSCAD draws anything circular. You can find more about how these variables impact your drawing in Chapter 8.

```
cylinder=(h=10,r=15);
```

Try setting $fn to very low values like 2, 4, or 5 and see what results. This can actually be useful for drawing triangles (3), hexagons (6), or octagons (8).

You can also create cones by setting the cylinder's r1 and r2 parameters:

```
cylinder(h=10,r1=15,r2=0);
```

The closer r1 and r2 are, the more the cone will look truncated. Try changing r2 to 12, for example. Don't forget that you can add center=true to any of your cylinders if you want the origin to run through the center instead of the edge.

These are all the parameters the cylinder command recognizes:

- h – Height (default=1)
- r – Radius of both ends (default=1)
- r1 – Radius of cone bottom (default=1)
- r2 – Radius of cone top (default=1)
- d – Diameter version of r
- d1 - Diameter version of r1
- d2 – Diameter version of r2

- center – Set to true to center cylinder or cone

Note you can specify a diameter (d) or a radius (r), whichever is more convenient.

OpenSCAD can also draw spheres using the sphere command:

```
sphere(r=10);
```

You can specify a diameter (d) instead of a radius (r), if you prefer. The $fn variable applies just like it did to cylinders.

The Polyhedron

Given rectangles (cubes), cylinders, cones, and spheres you can construct almost anything via a process known as solid constructive geometry. That's a fancy term for adding and subtracting shapes from either other (along with a few other simple operations). That's the topic of the next chapter.

However, you also sometimes need to make complicated shapes that would be hard to construct out of simple shapes. For that, OpenSCAD provides the polyhedron command. It is a little more complicated than the other primitive shapes covered in this Chapter. You don't often need them, so you may want to skip this section on a first reading.

A polyhedron takes two important parameters: points and triangles. The points parameter is a list of X,Y,Z coordinates. These will be the vertices of the polyhedron object. For example, consider a pyramid shape. There will be four points, one at each corner of the base plus one point up at the top. For example:

```
pyr_points=[ [25,25,0],[25,0,0],

    [0,0,0],[0,25,0],

    /* apex */ [12.5,12.5,25] ];
```

You can think of each of these points as having a number starting with zero. So [25,25,0] is point 0. The apex (or top) point of [12.5,12.5,25] is point 4.

The triangles parameter is another list of triples but these are not X, Y, Z coordinates. Instead they are the points from the points list that form triangles. Here's the rest of the pyramid:

```
pyr_triangles=[ [0,1,4], [1,2,4],

   [2,3,4],[3,0,4],[1,0,3],[2,1,3]];

polyhedron(pyr_points,pyr_triangles);
```

This looks complicated, but really all it is saying is that there is a triangle formed by point 0, point 1, and the apex (point 4). There is another triangle formed by points 1, 2, and the apex. The last two points form the base (as two triangles). Try removing the last two points and press F5. Then use the mouse to look under the pyramid. Put them back and observe the difference. You can also activate the View | Show Edges menu to see the triangle edges if you like.

Many 3D drawing systems use triangular tessellation (that is, they decompose all objects into triangles). Even a sphere can be built with little tiny triangles, each fitting together like a jigsaw puzzle. However, you would find it difficult to describe a sphere using OpenSCAD's polyhedron command. Luckily, you don't have to.

In Summary

In this chapter, you've learned a few simple OpenSCAD commands. They may not seem very exciting now, but they are the building blocks that will let you create larger and more complex designs.

Chapter 3. Operations

In the last chapter you made boxes and cylinders and spheres. You can use these as basic building blocks to create other things. To do that, you need a few more commands at your disposal. First, you need a way to move the objects you create around. Sure, the can center something or not, but you can't really move it around or rotate it. The other thing needed is a way to mix objects together. The simplest way to do that is with a union which glues two parts together. But you can also take the difference of two objects which removes the volume of one from another. So a cube differenced with a cylinder will look like a box with a hole drilled in it. There are a few other operations (such as intersection). Armed with these tools and a few basic shapes you can make a lot of different designs.

First, let's move things around. Moving something in OpenSCAD is a translate:

```
translate([0,25,0]) cube(15);
```

This moves the cube 25 units in the Y direction. If it doesn't seem to work, make sure you put the square brackets in the code. OpenSCAD doesn't complain if you put extra things in a call like translate. So if you write:

```
translate(0,25,0) cube(15);
```

OpenSCAD is happy to oblige and since the first parameter is zero, the translation does nothing. In theory, the parameter name for translate is v, but you rarely see it written out since it is the only parameter.

Transformations like translate apply to the next thing (in this case a cube). That's why it doesn't have a semicolon afterwards. However, sometimes you want to apply a transform to more than one thing. You can use curly braces to show what applies:

```
translate([0,25,0]) {
    cube(15);
```

```
        cylinder(h=20,r=25);

}
        sphere(r=6);    // not translated
```

It is customary to refer to the entire drawing as a "tree" because of the internal representation that OpenSCAD uses. For that reason you will sometimes hear people talk about a "subtree" when referring to different parts of the drawing. So in the example, the translate transform applies to one subtree (consisting of the cube and cylinder) while the sphere is another, untranslated, subtree.

Another common transform is the rotate transform. This rotates things (in degrees) around the X, Y, and Z axis. Try this line first:

```
cylinder(h=20,r1=25,r2=0);
```

Now try:

```
rotate([0,45,0]) cylinder(h=20,r1=25,r2=0);
```

You can also specify an angle (a) and an axis vector v. So to rotate 90 degrees around both X and Y axis you could write:

```
rotate(a=90, v=[1,1,0]) cylinder(h=20,r1=25,r2=0);
```

You can combine transforms too:

```
translate([-10.5, 15, 25]) rotate([0,45,0])
cylinder(h=20,r1=25,r2=0);
```

As long as you are doing easy transforms, try this:

```
color("red") translate([-10.5, 15, 25]) rotate([0,45,0])
cylinder(h=20,r1=25,r2=0);
```

You can also specify color as an RGB (red/green/blue) triple. So to make yellow (equal parts green and red) try changing color("red") to color([.5,.5,0]). Each color's value goes up to 1 so the example has half intensity green and red. If you add a 4th parameters (or an extra parameter when using a name) that sets a transparency (known as an alpha). An alpha of 1 is not transparent at all and zero makes the color totally transparent.

Keep in mind, though, that the colors only show up on a preview (F5). If you do the full render (F6) you won't see any colors because the CGAL rendering engine doesn't support color. That's the engine that you will use to generate an STL file, so you won't have multiple colors there either. Some transforms can also cause you to lose colors. Unless you are using them just for display (for example, to understand how some parts are fitting together) you probably won't use colors.

Start with an empty document and try this:

```
union() {
    translate([-10,0,0]) sphere(r=5);
    rotate([0,90,0]) cylinder(r=2,h=20,center=true);
}
```

Union glues the things in the curly braces together. This looks like two thirds of a barbell. Can you add the remaining weight?

The line you need to add (inside the curly braces) is:

```
translate([10,0,0]) sphere(r=5);
```

The cylinder is 20 units long and each weight is 10 units away from the center. Sometimes it is hard to see how the component pieces are going together in a union (or other operation). If you prefix the command (like the rotate before cylinder) with a # character it will highlight the shape in question on the output. Try that, by putting a # in front of the line that starts with rotate. If you want to see everything, use the View menu to select Thrown Together and press F5. This shows the cut out shapes in a different color.

19

Another handy character prefix is the % character. This makes the next object semi transparent so you can see inside of it. For example, try the barbell example with % before the translate lines that go with the sphere commands. That is:

```
union() {

    %translate([-10,0,0]) sphere(r=5);

    rotate([0,90,0]) cylinder(r=2,h=20,center=true);

    %translate([10,0,0]) sphere(r=5);

}
```

The result appears in Figure 3-1. Instead of a # or a %, you can use the ! character if you want to only draw that one thing. This (presumably temporarily) stops doing everything else, but shows you the current object you want to visualize.

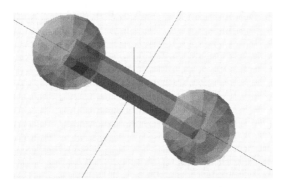

Figure 3-1. Transparent Objects with the % Prefix

Differencing

The other key operation is the difference operator. This draws the first thing you provide and subtracts off all the other things. Consider the modified bar bells in Figure 3-2. Here's one way to describe them in OpenSCAD:

```
union() {
// one weight
   translate([-10,0,0])
      difference() {  // cut weight in half
         sphere(r=5);
         translate([0,-5,-5]) cube(10);
      }
// bar
   rotate([0,90,0]) cylinder(r=2,h=20,center=true);
// other weight
   translate([10,0,0])
      difference() {  // cut weight in half in the other direction
         sphere(r=5);
            translate([-10,-5,-5]) cube(10);
      }
}
```

This is a little more complicated, but it is simple if you look at it one piece at a time. The union is what will get all three pieces (the two weights and the bar) together. The first translate is one of the weights. The sphere is around the center origin point and a box will, by default, cut a quarter of it out since it will have one corner at the center of the sphere. The translate slides the box over so it is cutting out the half of the weight we want.

If you want to better visualize this, try just these two lines:

```
   sphere(r=5);
   %  translate([0,-5,-5]) cube(10);
```

You can easily see how the cube will cut half of the sphere away.

The bar is easy. It is a simple rotated cylinder. Since it is centered, the weights will be at -10 and 10 on the X axis (since the bar is 20 units long).

The next translate is for the other weight. This uses the same logic as the first weight but has to slide the box to a different position to cut off the opposite hemisphere. Try changing the translate line in the two line example above to see how it moves the box.

So even though it looks intimidating, the program is actually pretty simple. Soon, you'll learn how to make programs like this more readable, as well.

Scaling and Resizing

There are a few other basic transforms you should know. The scale() transform stretches or compresses an object in any dimension. For example, if you take the barbell program and put as the first line:

```
scale([.5,2,1])
```

The barbell will be half as large on the X axis, but twice as large on the Y axis. The Z axis won't change at all. You can use resize, which is similar but makes the drawing fit in the size requested. For example, try resize([50,5,5]) instead of the scale command to see a longer stretched out barbell. If you set the auto=true parameter, any zero scale will be automatically scaled in proportion to the other scaling. For example, resize([50,0,0],auto=true) will stretch the barbell to 50 on the X axis but keep the other proportions the same.

Mirroring

Another less common transform is mirror(). As the name implies, it mirrors the object around the axis you specify. Here's the barbell example with one weight colored red and the other green. If you comment the mirror statement out, you'll see what effect it has.

```
mirror([1,0,0])
```

```
union() {

 translate([-10,0,0])

    difference() {

      color("red") sphere(r=5);

         translate([0,-5,-5]) cube(10);

    }


  rotate([0,90,0]) cylinder(r=2,h=20,center=true);

  translate([10,0,0])

    difference() {

      color("green") sphere(r=5);

         translate([-10,-5,-5]) cube(10);

    }

}
```

Intersection

Just as union takes multiple objects and merges them, intersection gets rid of anywhere the objects don't overlap. Here's an example:

```
intersection() {

sphere(12);

translate([10,0,0]) sphere(15);

}
```

Figure 3-2 Shows the two spheres without the intersection and Figure 3-3 shows the result after the intersection transform.

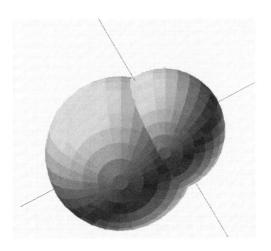

Figure 3-2. Two Spheres

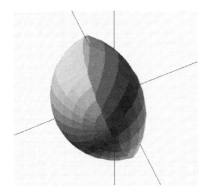

Figure 3-3. The Intersection of the Two Spheres in Figure 3-2

Math and Variables

There are many math operators supported by OpenSCAD that might be useful when defining objects, especially when used with variables. Unlike most programming languages, however, variables get set at compile time so (in general) the last thing you set the variable to will be used throughout the program.

If you aren't a programmer, this might not seem confusing, but if you are it will seem strange. It means statements like x=x+1 make no sense in OpenSCAD. Neither does something like if (x==0) x=5. You'll learn more about variable scoping in a future Chapter, but for now consider this example, which uses the echo command to print to the console:

```
x=5;

echo(x);

x=10;

echo(x);
```

The console shows:

ECHO: 10

ECHO: 10

It also shows an error message because I didn't ask it to draw anything. So, in OpenSCAD, variables are more like constants that you can set to something. Variables you don't set get the special value "undef" which is not often useful.

OpenSCAD supports basic math on numbers. That is + and – add and subtract. The * operator multiplies two numbers and / does division. Math is performed using the standard order of operations (that is, 5+3*2 does the multiply first so the answer is 11 not 16). If you want to force a different order, use parenthesis. So (5+3)*2 is equal to 16 and 5+(3*2) is unnecessary, but legal, and equals 11. Along with division, there is also a % operator which is remainder from integer division and is treated like division. So 10%3 is 1 since 3 goes into 10 but leaves a remainder of 1.

The + and – operators can also operate on two vectors (the lists of numbers between square brackets). Each element gets added or subtracted with the corresponding element of the other vector. You can also use * on two vectors which generates a single number: the dot product of the two vectors. If you have a matrix (like [[1, 2, 3], [4,5,6]]) you can multiply it by a vector or another matrix with the * operator.

A lot of drawings will make use of trigonometry. OpenSCAD operates in degrees (360 degrees in a full circle) and provides cos, sin, tan functions along with their inverses (acos, asin, and atan). There is also a two-argument atan named atan2 that can return the correct sign of the angle based on information in the second argument.

Here are some other common math functions:

- abs – Returns the absolute value of a number (e.g., 5 and -5 both return 5).

- ceil – Returns an integer by rounding up. That is, ceil(3.141) is 4 but ceil(-10.2) is -10.

- exp – Raise the natural base e to a specified power.

- floor – Returns an integer by rounding down. That is, floor(3.141) is 3 but floor(-10.2) is -11.

- ln – Natural logarithm.

- len – Returns the length of a vector, a matrix, or a string.

- log – Base 10 logarithm.

- lookup – Takes two argument, a value and a matrix (really a vector of vectors). Each vector has two elements, a key and a value. The closest match of the first value to a key causes lookup to return the associated value.

That one is easier to understand by example:

```
x2=lookup(3,[ [1,2], [2,4], [3,6], [4,8]]);
```

This sets x2 to 6 since the matching vector is [3,6]. If there is no exact match, OpenSCAD does linear interpolation. So changing 3 to 3.14 results in 6.28, since OpenSCAD realizes it must be between 6 and 8 by the same proportion that 3.14 is away from 3 and 4.

- max – Returns the largest of two values.

- min – Returns the smallest of two values.

- pow – Raises the first number to the second power. So pos(10,2) is 10 to the second power or 10 squared. Keep in mind that raising to .5 is the same as a square root, raising to 1/3 is a cube root, etc. Also see sqrt().

- rands – Generates a random number. Takes a minimum value, a maximum value, the number of values you want in the returned vector, and an optional seed value (if you start with the same seed you will get the same random numbers). Note the result is always a vector and the results do not have to be integer numbers.

Here's an example:

```
q=rands(1,10,1);
echo(q);
```

If you press F5 repeatedly, you'll see the ECHO message in the console reports a different number each time (for example, I just got 5.76679; the result will always be between 1 and 10).

- round – Rounds a number by the usual rules (that is, if the fractional part is less than 0.5 round down, otherwise round up).

- sign – Returns -1 for negative numbers, 0 for zero, and 1 for positive numbers.

- sqrt – The square root of a number (also see pow).

Printing

The barbells wouldn't make a good 3D print because they have no flat surface to rest on the printer bed. However, with the tools you have now, you can easily make complex objects by composing them of simpler objects. Here's a small stand with a letter X on it (or it could be a plus sign if you twist it around some):

```
union() {
cylinder(h=50.8,r1=50.8,r2=40);
rotate([0,0,45]) translate([-30,0,50.8]) cube([60,5,4]);
rotate([0,0,-45]) translate([-30,0,50.8]) cube([60,5,4]);
```

}

To print this object, you'd need an STL file. Press F6 to compile the drawing and then use Design | Export to STL on the menu. There are other available file formats, but for 3D printing you'll almost always use STL.

You need to compile first, because the default rendering is fast but does not generate an accurate enough representation of your model. Another concern is that your objects are manifold and watertight. That's a fancy way of saying a solid object has to be solid with no holes that open to the interior. That does not mean you can't have a tube or a plate with holes in it. It means the holes have to be constructed so that they have exterior surfaces all around the hole. This is called "watertight" because if you could fill the model with water and none would pass exterior surfaces the model would not leak.

Another important characteristic of objects it they do not have coincident faces. That's why when cutting an object out of another object (a difference transform) you have to make the cutting object big enough to extend past the edge of the main object. If the two faces are exactly the same, OpenSCAD will show a shaded color and you will have trouble creating a printable object.

For example consider this code:

```
difference() {
  cube(10,center=true);
  cylinder(r=2,h=10,center=true);
}
```

If you view this in OpenSCAD you will see the cylinder isn't fully cut out, but has a green shaded faces (for the default color scheme) that indicates a coincident face. You need to make the cutting cylinder a little bigger where it will "hit the edge" of the cube. Since that area is outside the cube, you could even make it ridiculously bigger. You may also need to translate it to make sure it cuts through all the faces. In this case, making the cylinder height 100 easily cuts through the whole cube (or you could set it to 11 if you prefer). You can visualize this with Thrown Together mode (on the View menu) or with the prefix characters discussed earlier.

Of course, your printer and slicing software may have other limitations about total size, resolution, and ability to handle overhangs, bridges, and other difficult to print features. Depending on many other factors, you may have to enable printing support to print difficult objects. None of these are particular to OpenSCAD, however, and depend on your printer and software choices.

A Simple Panel

Suppose you want to make a simple front panel for an electronic gadget. The panel is 125mm wide by 50mm high and should be 2mm thick. There are four 5mm holes for LEDs centered on the panel. There are also 4 #6 bolt holes for mounting 8 mm from each corner.

It is easy enough to make a blank panel by entering the following language into the top left pane:

```
cube([125,50,2]);
```

Enter that and press F5 (or use the Compile item from the Design menu). The result should look like Figure 3-4. You can use your mouse to change the view. Left dragging rotates the view. Right dragging pans. You can use the scroll wheel to zoom in and out.

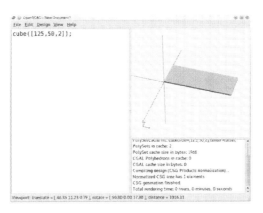

Figure 3-4. A Simple Panel

Just a blank panel isn't very exciting though. Also, since the numbers are hard coded, you'll have trouble later if you want to change anything. Let's fix that first:

```
panelWidth=125;

panelHeight=50;

panelThick=2;

cube([panelWidth,panelHeight,panelThick]);
```

This changes nothing, but it makes it much more readable. Later, you'll use these variables in other places and it will make things simpler to change. For example, let's add the LED holes. First, add the following line to the bottom of the ones that are already there:

```
cylinder(h=panelThick*2,r=5,center=true);
```

When you press F5 again, you'll be disappointed. The cylinder is right at the corner of the panel. Not to mention, you need a hole and this is another solid. We can fix the first problem by translating the cylinder to a new position. Change the cylinder line so that it reads:

```
translate([panelWidth/5,panelHeight/2,panelThick/2]) {

cylinder(h=panelThick*2,r=5, center=true);

}
```

When you press F5 now, you'll see the cylinder has moved to where the first LED hole should be. You may have to tilt the view a bit with the mouse to see it. Because it is tied to the variables you set up earlier, if you change the size of the panel, the hole will move to the right spot. The problem is, it isn't a hole, it is a cylinder poking out of both sides of the panel (thanks to moving the Z axis to panelThick/2). That's easy to fix with a difference operation. Here's the entire file (so far):

```
panelWidth=125;
panelHeight=50;
panelThick=2;
difference() {
cube([panelWidth,panelHeight,panelThick]);
  translate([panelWidth/5,panelHeight/2,panelThick/2]) {
  cylinder(h=panelThick*2,r=5, center=true);
  }
}
```

Of course, this is just one hole and you need 4. That's easy to add with a for loop (the subject of another Chapter):

```
panelWidth=125;
panelHeight=50;
panelThick=2;
difference() {
cube([panelWidth,panelHeight,panelThick]);
  for (i=[1:4]) {
    translate([panelWidth/5*i,panelHeight/2,panelThick/2]) {
    cylinder(h=panelThick*2,r=5, center=true);
    }
  }
}
```

The bolt holes at the edge are similar. A #6 bolt hole has a clearance hole of 0.1495 inches and there are 25.4mm in an inch. Some more cylinder commands will do the trick:

```
panelWidth=125;

panelHeight=50;

panelThick=2;


difference() {
cube([panelWidth,panelHeight,panelThick]);
  for (i=[1:4]) {
    translate([panelWidth/5*i,panelHeight/2,panelThick/2]) {
    cylinder(h=panelThick*2,r=5, center=true);
    }
  }
// bolt holes #6
  translate([8,8,panelThick/2])
    cylinder(h=panelThick*2,r=0.1495*25.4,center=true);


  translate([8,panelHeight-8,panelThick/2])
    cylinder(h=panelThick*2,r=0.1495*25.4,center=true);


  translate([panelWidth-8,8,panelThick/2])
    cylinder(h=panelThick*2,r=0.1495*25.4,center=true);


  translate([panelWidth-8,panelHeight-8,panelThick/2])
    cylinder(h=panelThick*2,r=0.1495*25.4,center=true);

}
```

Figure 3-5 shows the finished design.

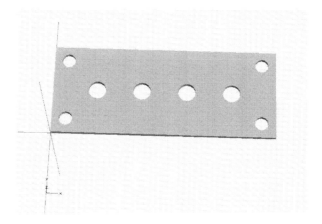

Figure 3-5. The Finished Panel

Summary

In this chapter, you've created simple objects by combining—in both a positive and negative way—basic shapes. There's a lot more you can do with OpenSCAD, but if you didn't go any further, you'd find you actually accomplish quite a bit with just the tools you have so far.

Chapter 4. Two Dimensions and Extrusions

Since you are interested in 3D printing, you might wonder why you would care about two dimensional objects in OpenSCAD. However, you can use 2D objects to create 3D objects and sometimes that is the best way to go.

The 3D shapes you have already used all have 2D matches. A square instead of a cube, a circle instead of a sphere or cylinder, and a polygon instead of a polyhedron. Try this:

```
square([12,5]);
```

Not very exciting. The trick is that you can use that square to create different 3D shapes. The easiest way is to simple extrude it using linear_extrude. In manufacturing, extrusion is pushing something through a shape. So a circular shape makes a wire. A "U" shape might make an aluminum channel. For OpenSCAD, the idea is the same. The 2D shape is like an extrusion die and the solid is what you would get if you pushed material out of that 2D shape.

Try this:

```
linear_extrude(height=20) square([12,5]);
```

Of course, you could just have easily done this with a cube. You can apply transforms to 2D shapes, too:

```
linear_extrude(height=20)
difference() {
square([12,5],center=true);
circle(2);
}
```

Naturally, you could this with a cube and a cylinder, just as easily. However, there are other options when you extrude something. For example, you can apply a scaling factor or you can rotate (twist) the object as it is extruded. If fact, if you want to, you can do both at the same time.

Consider scaling. As the extrusion stretches out the 2D shape, it starts at a scale of 1:1 and then makes it so at the end it will be at the specified scaling factor. In between, it will change the scale in a linear fashion.

Think about how hard it would be to make a pyramid with a bunch of cubes. You'd have to rotate several cubes and subtract them from the main cube. However, if you start with a square, extrude it, and scale it to zero, you get a nice pyramid:

```
linear_extrude(height=20,scale=0)
    square(12,center=true);
```

Notice the scale reduces to zero over the length of the object. You could grow the scale, instead. Try changing scale to 10. You can use a vector of two elements to scale asymmetrically, if you want. Try setting scale to [2,10].

You can also cause an extrusion to twist through any number of degrees. Since 360 degrees is a full twist, using more than that implies you want more than one full twist. For example, try setting the twist to 45, 90, 270, 360, and 720 in the following example:

```
linear_extrude(height=20,twist=45) square(12,center=true);
```

If you want a very unique pyramid, you can try a little of both:

```
linear_extrude(height=20,twist=125, scale=0)
square(12,center=true);
```

The result appears in Figure 4-1. If you want to make it appear smoother, try adding slices=100 to the linear_extrude command. You can also specify $fn and the other circle control variables.

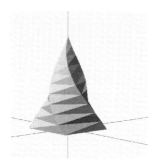

Figure 4-1. A Unique Pyramid

The linear in linear_extrude implies that it stretches the 2D object in a straight line (disregarding twist and scale). However, you can also take a 2D object and rotate it to create a 3D object.

Try this:

```
rotate_extrude($fn=100) translate([2,0,0]) circle(1,$fn=100);
```

That makes a donut. If you didn't translate the original circle, you'd get a sphere.

Extruding polygons can make very sophisticated shapes. For example, consider Figure 4-2. It was created with this command:

```
rotate_extrude($fn=200) polygon(
points=[[0,0],[2,1],[1,8],[1,9],[3,11],[5,15]] );
```

If you want to see the cross section, just run the polygon command alone and rotate the view.

Figure 4-2. An Extruded Polygon

Extruding Imports

Another important item you can extrude is an imported 2D shape. You could create these in a separate program like Inkscape or a CAD program. Here's the basic command:

```
import (file = "example.dxf", layer = "layer_0", origin =center);
```

You specify the file name and the layer you want from that file. You can also pick an origin, although you don't have to.

Copying a Flat Solid Object

Here's an example of how I took a wooden letter "W" that had broken, took a photograph of it, digitized it, repaired it, and then make a 3D model out of it using the techniques in this Chapter. You can find the files on Thingiverse at http://www.thingiverse.com/thing:43387.

The first job was to get a picture. It didn't need to be a great picture, but it did help to get it on a high contrast background. I just used a simple digital camera to take the picture (see Figure 4-3). Notice the break near the bottom left (just as the stroke of the letter starts to go towards the central peak).

Figure 4-3. The Original Wooden Item (Complete with Break)

The problem, of course, is that the photograph is made of pixels, and the 3D printer wants a 3D model made up of triangles and vectors. The plan of attack goes like this:

- Use an image editor (GIMP) to convert the picture to a simple black and white image with clean edges (and the damage repaired, in this case).

- Use a vector image editor (Inkscape) to trace around the letter and produce an SVG file (a 2D vector format).

- Import that vector image into OpenSCAD as a 2D object. Use the linear_extrude command to "extrude" the 2D path into a 3D object.

- Print the STL file and presto, a new perfect W is born. The results appear in Figure 4-4 (a small test print and a full size print).

Figure 4-4. Two Printed Copies of the Original (at Different Scales)

Images from digital cameras are made of pixels (a raster image). Inkscape uses vectors—instead of dots, the image is described as lines and directions. This is more similar to the way the 3D STL file format works. That's why the conversion has to go through Inkscape and not directly into OpenSCAD.

GIMP

Before you can convert your image into a vector format, it needs to be as simple as possible. The shape should be all one color (usually black) and the background should be a highly contrasting color (like white). Like most photo editors, GIMP provides a variety of smart selection tools. I used the fuzzy selection tool (the one that looks like a magic wand) to select the object. If I missed some spots, I held down the shift key and added more to the selection. In the case where I got too much, I used the control key to subtract a new selection from the existing one.

Keep in mind that you can start with the magic wand tool and then use other selection tools to add or subtract. For example, adding a rectangular selection was useful for repairing the break in the letter. If there is any shadow or any part of the vertical faces of the original in the picture, don't select them. You want the selection to be a representation of the flat face of the part. This also means when shooting the picture you should try to get the camera lens as close to perpendicular to the part as possible. By the same token, taking the picture against a solid-colored contrasting background will be very helpful as well.

Once you have the selection outlined, you can press the delete key to replace the entire selection with a solid background color. By the color selector in the toolbox there are some swap arrows that let you swap the foreground and the background color. You can press then and then right click inside the selection and pick Select | Invert. This will cause everything that is not the part to become selected. Hit delete again to change this background to a solid color (the original foreground color).

GIMP's native file format isn't useful in this process, but you might want to save a copy anyway in case you want to touch up the selection process. The working file, however, needs to be a JPEG and to do that you have to use the File | Export menu item, not the File | Save command.

Once you have the JPEG you've done most of the hard work. Figure 4-5 shows the end result of the processing with GIMP.

Figure 4-5. The Photo Converted to Pure Black and White

Inkscape

When you launch Inkscape it will start with a blank page. The page size isn't going to make any difference in this case, so don't worry about it. Start by importing the JPEG file you saved from the first step. The goal is to get a path around the outline of the part. This ought to be hard, but Inkscape has a Path | Trace Bitmap command (see Figure 4-6) that will do the work automatically. Because of the work done in GIMP, the settings on the trace command aren't especially critical. You can play with them until you get a good looking path. You can use the Update button to get a preview or you can simply press OK and then undo anything that wasn't satisfactory.

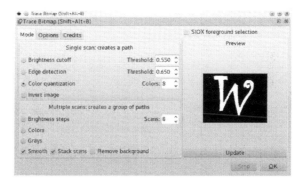

Figure 4-6. Inkscape's Trace Bitmap Dialog

Once you press OK and you are happy with the result, you can press the Edit Paths button (or press F2) and select your new path. You'll see lots of nodes lit up. You can delete any extraneous nodes (you can even select lots of them with a drag and delete them *en masse* if you like). You can also grab the nodes and pull or push them to modify the shape of the path. When you are satisfied you might want to try Path | Simplify to reduce the number of nodes.

You may find it easier to delete the JPEG under the path after tracing, or you may like to leave it there while you work for reference. However, before you save the file as a DXF (a standard CAD file format) you need to delete the bitmap and leave only the path.

In theory, you can save the file using Inkscape's "Desktop Cutting Plotter" format (you can find all the formats on the Save As dialog). However, the export in this format is a bit too simple, and most people prefer to use a third party add-on to save to DXF. One that works well is provided free from Big Blue Saw (http://www.bigbluesaw.com/saw/big-blue-saw-blog/general-updates/big-blue-saws-dxf-export-for-inkscape.html). You might also try the one at http://www.thingiverse.com/thing:25036 although recent versions of Linux have trouble running this one. However you do it, you need to save your work as a DXF. Again, you might also save a copy as an SVG (Inkscape's native format) in case you want to go back and touch it up again.

If you wanted a cookie-cutter effect (that is, just the outline of the letter or shape) you could copy and paste the path so that there were two copies. Change the fill color of one so you can tell them apart. Then drag the copy so that it is exactly on top of the original. Use the Path | Inset command to "shrink" the copy a few times until you have the thickness you want (you'll be able to see because of the different colors—the original color will be the edge.

Once it is how you want, select the original and then use the shift key and select the copy. Path | Difference will then subtract the copy from the original, leaving an outline suitable for export to make a cookie-cutter.

OpenSCAD

OpenSCAD can import DXF files. Earlier, you saw the linear_extrude command followed by an import command. Older versions used an argument to the linear_extrude command to specify the file. Newer versions complain that you ought to be using the new way, but still allows the older method. Just in case, here's both ways to do it.

With older versions of OpenSCAD, this was the line of code required to read the /tmp/w.dxf file:

```
linear_extrude(file="/tmp/w.dxf", height=5,center=true);
```

However, the "new" way to do it looks like this:

```
linear_extrude(height=5,center=true)
import(file="/tmp/w.dxf");
```

The result should be the same. The file is converted to an OpenSCAD geometry and then it is extruded 5 millimeters to be a 3D object. If the object shows up as an outline, you didn't get down to two colors in GIMP (that actually could work if you wanted a cookie cutter). For this project, the result should be a solid object as seen in Figure 4-7.

Figure 4-7. The Result in OpenSCAD

Once you have the line entered (setting your desired extrusion height and file name, of course), you simply press F5 or F6 (Compile or Compile and Render). Then you can execute the Design | Export as STL menu item to save your STL file.

Of course, you could just use the imported shape just like any other OpenSCAD object and build up something more complex. For example, consider this script which produces the output seen in Figure 4-8.

```
$fn=64;

union() {

cylinder(height=5,r=15);

translate([0.7,-8,0])

  linear_extrude(height=5,center=false)

    import(file="/tmp/w.dxf");

}
```

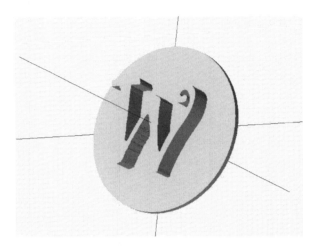

Figure 4-8. Adding Shapes in OpenSCAD

Print!

Armed with the STL file, you can print just like any other model. You might need to adjust the scaling a bit to get the size to exactly match the original (if that's important). The ability to copy real objects with a digital camera is pretty powerful. Of course, 3D scanners exist and there is even software that can attempt to take multiple pictures and change them to a 3D object. Those are outside the scope of this book, but still an exciting possibility.

It is worth noting that you don't have to start with a picture of a real object. The same technique works fine if you want to draw something in Inkscape and then print it. That's how the object in Figure 4-9 was created. The text started out as an Inkscape object. You must convert everything to a path and then the rest of the process was just the same as printing the giant letter.

Figure 4-9. Inkscape Text Converted to Plastic

Wrap Up

In this chapter, you've seen how to create 2D objects and promote them to solids using different kinds of extrusions. For many jobs, you'll never need this technique. However, for certain kinds of objects, it is easier to use extrusions than it is to model them totally with solid objects.

Chapter 5. Programming OpenSCAD

So far, you've mostly used commands to build and combine shapes. Much of the real power of OpenSCAD, though, comes from being able to handle programming tasks to make assemblies with regular features that adapt or customize to changing situations.

You've already seen that comments can make scripts more readable. The // comments run to the end of the line while /* comments run to the next occurrence of */ which could be on the same line or many lines away.

Variables

Variables can also make code easier to read, but they aren't exactly like variables in a regular programming language. The value of variables are set, for the most part, when the code is parsed not when it is executed. So although it seems strange, the following code prints 10 to the console twice (the echo command prints a value to the console):

```
x=20;

echo(x);

x=10;

echo(x);
```

Only the last x= statement counted. You'll see that this applies in most situations, but each module has its own values. In addition, the assign command can change a value in a specific way.

Modules and Functions

Modules let you collect bits of OpenSCAD code into reusable bits. This is efficient and it also promotes readability. Here's a simple module and function:

```
holer=2;

thick=6;

function half(n) = n/2;

module hole(x,y) {
// try this
// holer=20;
   translate([x,y,0]) cylinder(r=holer,height=thick);
}


cube([30,30,thick],center=true);
hole(10,half(20));
// this will override holer=2 above
// for both calls to hole
holer=6;
hole(-10,half(20));
```

If you remove the comment from the holer=20 line you'll see that the line only applies to that module. Otherwise, the final value of holer is 6 (from the last assignment).

Keep in mind that a module (or a function) all by itself does nothing. You have to call it (as the example calls "hole" in two places) for it to do anything.

You can specify default values for variables in modules and you supply them just like you supply values to built-in modules like cube:

```
module samplebox(w=10) {

   cube(w,center=true);

}

samplebox();   // 10

samplebox(20);

samplebox(w=20);
```

Interestingly, modules inherit any variables you set even if they aren't in the list. For example:

```
module foo(x) {

    echo(y);

}

foo(y=10);

foo(y=20);
```

Even though y isn't mentioned in the module's argument list, you can still set y on the call. Of course, you can't set a default in this case.

Another neat trick for modules is that you can use the "children" command to act on any thing that comes after the call to the module. In older version of OpenSCAD, the command is child instead of children. For example:

```
module move_to_x(x) {

// use child() if children doesn't work

   translate([x,5,0]) children();

}

move_to_x(10) sphere(5);
```

You could just as well use curly braces and have move_to_x apply to many children. Sometimes you want to apply things to different children. In that case, use children(0) (or child(0)) to refer to the first child, children(1) to refer to the second, and so on. The total number of children is available in the $children variable.

For Loops

The real motivation for using variables, modules, and functions is to make objects you can quickly change. Suppose you had a panel with some holes in it in a graphical CAD program. If you made the panel larger or smaller, you'd have to move each hole to the its new position. In OpenSCAD, you simply describe where the holes are using a formula instead of a constant, and the holes will reposition automatically. That assumes, of course, that you wrote a smart formula.

Consider this program:

```
width=25;

height=15;

thick=3;

nholes=5;


difference() {

  cube([width,height,thick]);

      for (i=[1:nholes]) {

    echo(i);

      translate([width/(nholes+1)*i,height/2,thick/2])
cylinder(r=holer,h=thick+3,center=true,$fn=25);

  }

}
```

The key is that each hole is translated to an X and Y position given by width/(nholes+1)*i and height/2. The i variable is from the for loop. There are several ways to write the for loop statement, and the one in this example causes OpenSCAD to repeat the code that belongs to the for loop starting with i=1 and continuing until it executes the code where i=nholes. So in the example, the code will run the translate/cylinder commands 5 times with i=1 the first time, then i=2, and so on until i=5.

The colon in the for statement is important. That tells OpenSCAD it is a range. If you just put a normal vector in, the loop just executes once for each item in the vector. So if you made a typo and used [1,5] instead of [1:5] you'd get two executions: one with i=1 and one with i=5. Ranges can go backwards, if you like, so [5:1] works (and sets i=5 first).

If a range has two colons in it, the middle number is an increment. So to go from 0 to 10 by twos, you'd write [0:2:10] and i would get set with 0, 2, 4, 6, 8, and 10.

If you use a vector, then the loop variable (which is often named i for historical reasons) will take on vector values. In that case, for one, you may need to nest two for loops together. You can do this by simply writing two for loops:

```
for (i=[0:2:10])
   for(j=[0,1]) echo (i+j);
```

Or you can specify them together:

```
for (i=[0:2:10], j=[0,1])
   echo (i+j);
```

You can add as many loop variables as you like (although there is probably some large limit somewhere, I haven't found it):

```
for (i=[0:2:10], j=[0,1], k=[0,0.1,0.2]) ...
```

Although you often see the loop variables written as i, j, and k, there's no reason they can't be any name you like:

```
for (length=[1:5], width=[1:5])
    translate([0,0, length]) cube([length,width,0.5]);
```

There is one wrinkle to using a for loop. By default, OpenSCAD does an implicit union of the shapes generated in a for loop. That makes it impossible to do an intersection of the elements in a for loop. To work around this, OpenSCAD provides intersection_for. This is just like a normal for, but does an intersection of each loop instead of a union.

Because variables don't get reinterpreted, things like this won't work:

```
for (length=[1:5], width=[1:5])
{
    index=width/5;  // will not work!
    translate([0,0, length]) cube([length,width,0.5]);
}
```

In fact, OpenSCAD is so sure you didn't mean to write this, it will throw an error.

However, you can use the assign statement, which sets a variable for the next statement (or group of statements inside curly braces). So the right way to write the above example is:

```
for (length=[1:5], width=[1:5])
  assign(index=width/5)
    translate([0,0, length]) cube([length,width,index]);
```

The index variable is computed each time in the loop and is only valid for the following statement (the translate/cube pair). If you try to print index out after the assign statement, OpenSCAD will tell you it is undef (undefined).

One classic programming construct that OpenSCAD provides is the if statement. This allows you to test a value and do something if the test is true. You can also optionally add an "else" part to do if the test is false.

You might use this to leave out certain things in a design based on a variable. For example:

```
w=25;

h=10;

t=3;

mtghole=1; // set to zero for no mounting holes

difference() {

// main panel

  cube([w,h,t]);

// switch cut outs

  translate([w/4,h/2,-.1]) cylinder(h=2*t,r=.5);

  translate([2*w/4,h/2,-.1]) cylinder(h=2*t,r=.5);

  translate([3*w/4,h/2,-.1]) cylinder(h=2*t,r=.5);

// mounting holes

  if (mtghole==1) {

    translate([.3,.3,-.1]) cylinder(h=2*t,r=.1);

    translate([w-.3,h-.3,-.1]) cylinder(h=2*t  ,r=.1   );

    translate([.3,h-.3,-.1]) cylinder(h=2*t,r=.1);

    translate([w-.3,.3,-.1]) cylinder(h=2*t,r=.1);

  }

}
```

Functions can't use if statements, but sometimes you need to alter the function behavior based on a value. For example, suppose you want a function that returns a number you pass into it unless the number is zero. If it is zero you want to return a default value. You can write:

```
function getvalue(value, default) = value!=0?value:default;
```

The general form is <test>?<value for true>:<value for false>. This is useful for recursive functions (functions that call themselves) that need to decide when to stop. Here's a factorial function, for example:

```
function factorial(x)=x==1?1:x*factorial(x-1);
```

You can use recursion in a module too:

```
module steps(z) {

  if (z<=50) {

    translate([0,0,1.1*z]) cube([z,z,z],center=true);

    steps(z+10);

  }

}


steps(5);
```

In either case, the variable gets set to a new value each time, a rarity for OpenSCAD.

The example panel above is actually a good candidate to use two features of OpenSCAD to make the code more readable and maintainable: modules and functions. You saw both earlier in this Chapter. Here's the same code rewritten to make use of modules, a function, an if statement, and even some for loops:

```
w=25;    // width of panel
h=10;    // height of panel
t=3;     // thickness of panel
nrswholes=3; // number of switch holes
mtghole=1; // set to zero for no mounting holes



function getwidth(n)=n*w/(nrswholes+1);


module panel() {  // main panel
    cube([w,h,t]);
}


module switch(n) { // switch holes
    translate([getwidth(n),h/2,-.1])
       cylinder(h=2*t,r=.5);
}


module mthole(x,y) {  // mounting holes
   translate([x,y,-.1]) cylinder(h=2*t,r=.1);
}


difference() {
// main panel
```

```
  panel();
// switch cut outs
  for (i=[1:nrswitchholes])
      switch(i);
// mounting holes
  if (mtghole==1) {
  for (i=[.3,w-.3], j=[.3,h-.3]) {
      mthole(i,j);
     }
  }
}
```

When using a difference() it is often useful to drop the hole a little bit and make it bigger than necessary since it ensure a clean cut all the way through the solid. That's why the cylinders are 2*t tall (which is more than necessary) and why they are all dropped -0.1 in the associated translations.

One important note about the if statement. As in many programming languages, the test for equality is two equal signs ("=="). If you use a single equal sign you will not get the result you expect. The operators available are:

- < - less than
- <= - less than or equal to
- > - greater than
- >= - greater than or equal to
- == - equal
- != - not equal

In addition, you can join expressions together with && (and) and || (or). You can reverse the sense of a test with the ! character. So for example:

> if (x<10 && y==2)   // if x is less than 10 and y is equal to 2
>
> if (!(x>3 && y>=0)) if it is not true that x>3 and y>=0

> Figure 4-9. Inkscape text converted to plastic

Thingiverse Customizer

When you share an OpenSCAD design on Thingiverse, you can have the web site customize your OpenSCAD file based on user input. The process is simple. Consider a plate with some #6 bolt holes in it. I use these to clamp belts, but you might use it anywhere you need a "mending plate." However, you might need different sizes or numbers of holes.

The trick is use variables with special comments. Here's the mending plate file:

```
/* Really simple "mending plate" with two #6 holes
   -- I use this to clamp a GT2 belt,
      but it probably could be used
   for anything. Al Williams, April 2013 */
// Length of plate (mm)
length=17; // [10:100]
// Width of plate (mm)
width=6; // [5:100]
// Thickness of plate (mm)
thickness=3; // [.5:20]
// Offset from edge/hole spacing (mm)
```

```
edgeoffset=3.5  ;   // [1:10]

// Number of holes (should be even)

n=2; // [[1:100]

loopct=n/2;

difference() {

cube([length,width,thickness]);

for (i=[1:loopct]) {

   translate([edgeoffset*i,width/2,1])

        cylinder(h=thickness*10,r=2,center=true,$fn=100);

   translate([length-edgeoffset*i,width/2,1])

        cylinder(h=thickness*10,r=2,center=true,$fn=100);

  }

}
```

    The web site will offer input for any variables that get a constant value. If you want "hidden" variables, you can add 0 to the constant to prevent the system from picking them up. In the example, loopct is the only variable that is hidden and it isn't set up from a constant anyway.

    Figure 5-1 shows an example of customizing the plate on the Thingiverse web site.

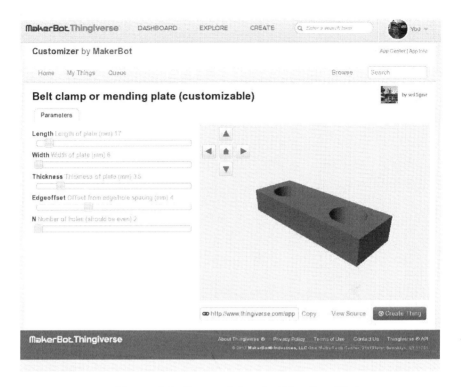

Figure 5-1. Customizing the Mending Plate

The comments directly before the variable set the explanatory text. The numbers in square brackets set the range of choices. If the range has a colon, the range turns into a slider (like the ones in Figure 5-1). If it is a comma-separated list, the customizer will create a drop down box. If you don't put any range at all, the customizer will use a regular input box. There are several other syntax tricks, but these will take care of most of your designs. You can read all about your choices at: http://www.makerbot.com/blog/2013/01/23/openscad-design-tips-how-to-make-a-customizable-thing/.

Wrap Up

In this chapter, you've learned how to use OpenSCAD as a sort of programming language. Because of limitations on variables and other restrictions it is probably more akin to a scripting language than a full-blown programming language. Still, careful use of programming features will let you have increased versatility, reusability, and maintainability of your 3D models.

Chapter 6. Libraries

In the last Chapter, you saw that you can make parts of your programs reusable by a judicious employment of variables, functions, and modules. The next logical step is to make those reusable pieces libraries that you (or anyone) can reuse without having to understand the underlying guts.

Any set of modules you create can form a library. OpenSCAD can load another file from your current program using one of two commands: include and use.

When you write:

```
include <lib/mylib.scad>;
```

OpenSCAD looks in the lib subdirectory of the current directory for a file named mylib.scad and executes it. The end effect is as if the file was just copied and pasted right at that point in the document. This statement:

```
use <lib/mylib.scad>;
```

Does nearly the same thing, but it ignores any commands in the library file. It only processes module definitions and function definitions. That's useful if you have a file that has modules you want, but you don't want it to render whatever it was that it was supposed to draw.

Depending on your operating system, OpenSCAD not only looks in the current directory for libraries, but also in some system-defined directories. There are many very useful libraries included in these predefined library directories. In addition, there are many libraries online including some very useful ones on Thingiverse.com.

An important note about variables in libraries: Remember that OpenSCAD variables get defined once (usually the last definition overrides all the others). Variables in a library are a little different. If you use include, a library could set up some variables. If you want to override them, you may, but you must do it before you include the library.

Some Predefined Libraries

By default, OpenSCAD installs the MCAD libraries so that you can use them. You can find the entire library on GitHub (https://github.com/SolidCode/MCAD). The documentation for each file is inside the source code, so looking at the GitHub source is almost not optional.

As a simple example, consider the nuts_and_bolts.scad file. This file has modules for metric bolt holes and nut holes:

```
use <MCAD/nuts_and_bolts.scad>;


difference() {

  cube([20,20,10]);

  translate([10,10,0]) boltHole(size=6,length=10);

}


difference() {

  translate([40,0,0]) cube([20,20,10]);

  translate([50,10,-2]) nutHole(6);

}
```

There are many useful MCAD libraries. Here's one that builds rounded boxes easily:

```
use <MCAD/boxes.scad>

roundedBox([30, 15, 40], 5);
```

A very useful third-party library is the write.scad library available on Thingiverse: http://www.thingiverse.com/thing:16193. This library lets you add text to your drawings easily using several predefined fonts. Follow the instructions to install the library and try this:

```
use <write.scad>


translate([10,10,0])

write("hello",h=20,t=5);
```

There are other modules that write text in special ways. For example to make curved text (see Figure 6-1), try:

```
use <write/write.scad>


cylinder(r=20,h=10,center=true);

writecylinder(

   "3D Printing",[0,0,0],

   20,10,face="top",center=true);

writecylinder("Coin",[0,0,0],-10,10,

   face="top",center=true);
```

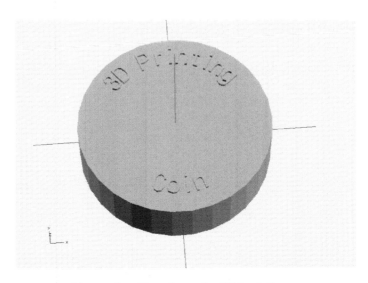

Figure 6-1. Text from the Write Library

By the way, the built-in MCAD library supports text through the fonts.scad library. It is a bit harder to use (in my opinion) but if you want to try it, pick example 023 from the File | Examples menu.

Another useful library puts a view-only plane under your drawing that represents your printer's build platform. This is handy to see when you create an object that is too big to print or if you want to layout multiple objects.

Download the library from http://thingiverse.com/download:121626 and install according to the instructions. Then customize the following and either put it at the top of your program, or make your own library (like my_build_plate.scad) with the contents:

```
use <build_plate.scad>

//set your build plate

//[0:Replicator 2,1: Replicator,2:Thingomatic,3:Manual]

build_plate_selector = 3;

//if set to "manual" set the  build plate x dimension

build_plate_manual_x = 200; //[100:400]
```

```
//if set to "manual" set the  build plate y dimension
build_plate_manual_y = 200; //[100:400]


build_plate(build_plate_selector,
  build_plate_manual_x,build_plate_manual_y);
```

Wrap Up

In this chapter, you've learned how to use OpenSCAD libraries to reuse your code and the code other people write for you. Without libraries, doing things like text would be very tedious. But as your toolbox grows both from outside sources and from your own unique tools, you'll find making new drawings gets easier and easier.

Thingiverse (as well as other Web sites) has many libraries you'll find useful. There are also tools that aren't really libraries like the polygon generator at http://www.thingiverse.com/thing:9290. Search for OpenSCAD to find many more libraries and tools.

Chapter 7. Advanced Topics

You could stop here and possibly never need to know any more about OpenSCAD than you have already learned in the previous Chapters. But there are a few more things that you might find useful as you try to do more sophisticated drawings.

Strings

You don't often need them, but OpenSCAD has some facility for dealing with strings of characters. The str() function converts any variable or constant into a string. One nonobvious use of this is to test if a variable contains a vector. You might write:

```
if (str(v)[0]=="[")
```

The [0] selects the first character of the string.

There is also a search function that allows you to find one string in another string. The basic use is you provide a search value and a target string (or vector). The return is an index into the target string.

Importing STL

Sometimes you have an existing STL model from another tool (or, perhaps, off the Internet) and you want to do something with it in OpenSCAD. Maybe you want to add to it, or cut part of it off. You might just want to scale or rotate it. You can do it by importing the STL file and manipulating it like any other geometric element.

Newer versions of OpenSCAD use the import command to load STL files, although older ones used import_stl. You could write, for example:

```
rotate([0,90,45]) import("my3dmodel.stl");
```

It isn't very useful for actual design, but if you want to show off your OpenSCAD creation, you might consider animating it. Select View | Animate from the menu and you'll see several boxes at the bottom of the window that show a time, a Frames Per Second (FPS) and a number of steps. You can change the FPS and steps and it will appear that nothing happens. That's because you need to make your model do something different at each time, which is represented by the $t special variable. For example, here's a rotating snowman:

```
// snowman




basesize=20;


module snowman() {

sphere(basesize);

translate([0,0,(basesize-5)*2+3]) sphere(basesize-5);

translate([0,0,(basesize-5)*3+11]) sphere(basesize-10);

translate([0,0,(basesize-5)*4+6]) cylinder(r=12,h=2);

translate([0,0,(basesize-5)*4+8]) cylinder(r=8,h=5);

}


rotate([0,$t*360,$t*-360]) snowman();
```

Advanced Transforms

You saw the common transforms like rotate and scale earlier, but there are two more transforms you may need for advanced modeling. One is the convex hull transform (hull). You can think of this transform and putting a rubber band around some points on a plane. You can read more about the math behind this at http://en.wikipedia.org/wiki/Convex_hull.

Another advanced transform is the Minowski or Minowski sum transform (see http://en.wikipedia.org/wiki/Minkowski_addition). This allows you to merge two shapes into one. Perhaps the best example is to create a box with rounded corners. The obvious way to do this is create a cube and several cylinders and union them together.

However, a more elegant way is to use the Minowski sum operation:

```
minkowski()
{
  cube([10,10,10]);
  cylinder(r=1.5,h=1);
}
```

This merges the two shapes so that each corner of the box is actually the cylinder.

You could do the same thing with the hull transform, by the way. In this case, you'd draw the corner cylinders in the right place and then use hull to put a "rubber band" around them forming the box. Since hull operates on a plane, you also need a linear_extrude to convert it back to a solid.

Of course, the Minowski is more elegant and even simply doing a union between the cylinders and a cube might be a little simpler. Here's the code:

```
x=10;
y=10;
z=10;
radius=1.5;


linear_extrude(height=z)
hull()
```

```
{
translate([(-x/2)+(radius/2), (-y/2)+(radius/2), 0])
circle(r=radius);

translate([(x/2)-(radius/2), (-y/2)+(radius/2), 0])
circle(r=radius);

translate([(-x/2)+(radius/2), (y/2)-(radius/2), 0])
circle(r=radius);

translate([(x/2)-(radius/2), (y/2)-(radius/2), 0])
circle(r=radius);
}
```

Projection

Sometimes, you may want to convert a 3D shape into some 2D representation. That's the purpose of the projection command. If you set cut=true, you essentially get a cross section of the affected object at z=0. Of course, you can rotate and translate the object beforehand to get a cross section at any position.

If you set cut=false you get what amounts to a shadow of the object instead of a true cross section.

Consider this example object:

```
difference() {
    sphere(20);
```

```
            cylinder(r=5,h=40,center=true);

        translate([0,0,20])
cylinder(r=12,h=15,center=true);

    }
```

To get a cross section through the middle, simply add this line to the top:

```
projection(cut=true)
```

If you wanted a cross section nearer the top you could write:

```
projection(cut=true)

        translate([0,0,-15]) difference() {

    sphere(20);

            cylinder(r=5,h=40,center=true);

        translate([0,0,20])

            cylinder(r=12,h=15,center=true);

    }
```

Wrap Up and Summary

In this chapter, you've learned a bit about some more advanced features of OpenSCAD. If you really want to dig into these topics, you'll want to read up on some 3D modeling theory (Wikipedia isn't a bad place to start) and read up on the OpenSCAD manual.

This book gives you the basic tools to use OpenSCAD to create 3D models that you can use for printing. With practice, you'll start to notice that everything around you is composed of just a few basic shapes and modeling will be much easier.

The Internet is also a wealth of libraries and examples. Keep in mind libraries are just text files you can read and you can learn a lot dissecting a downloaded library.

Thanks

Thanks for reading this book. Be sure to look for *Understanding 3D Printing* on the Amazon Kindle store.

Volume I. Understanding 3D Printing: 3D Printing Basics

http://www.amazon.com/Understanding-3D-Printing-Basics-ebook/dp/B00DS8RASG

Volume II. Understanding 3D Printing: About 3D Printing Hardware

http://www.amazon.com/Understanding-3D-Printing-Hardware-ebook/dp/B00DSCR92E

Volume III. Understanding 3D Printing: About 3D Printing Software

http://www.amazon.com/Understanding-3D-Printing-Software-ebook/dp/B00DTGB1YQ

Volume IV: Understanding 3D Printing: Refining your Prints

http://www.amazon.com/Understanding-3D-Printing-Refining-ebook/dp/B00DX4GC9I/

All volumes together

http://www.amazon.com/Understanding-Printing-Volumes-1-4-ebook/dp/B00DXHBD2G

Follow us:

http://www.facebook.com/Understanding3dPrinting

http://gplus.to/Understanding3D

Twitter: @awce_com

Made in the USA
San Bernardino, CA
01 April 2019